SCIENCE ON THE INTERNET

A STUDENT'S GUIDE

1999

ANDREW T. STULL

PRENTICE HALL Upper Saddle River, NJ 07458

Editorial Director: *Tim Bozik*
Editor in Chief/Science: *Paul Corey*
Editor in Chief/Development: *Ray Mullaney*
Ass't. Vice President/Production & Manufacturing: *David W. Riccardi*
Special Projects Manager: *Barbara A. Murray*
Manufacturing Manager: *Trudy Pisciotti*
Manufacturing Buyer: *Ben Smith*
Supplement Cover Manager/Designer/Illustrator: *Paul Gourhan*

TRADEMARK INFORMATION:
Microsoft Windows is a trademark of Microsoft Corporation; Netscape is a registered trademark of
Netscape Communications Corporation; Yahoo! is a registered tradmark of Yahoo Corporation;
Java is a registered trademark of Sun Microsystems

The authors and publisher of this manual have used their best efforts in preparing this book. The
authors and publisher make no warranty of any kind, expressed or implied, with regard to these programs or the documentation contained in this book. The author and publisher shall not be liable in
any event for incidental or consequential damages in connection with, or arising out of, the furnish-
ing, performance, or use of the programs described in this book.

Printed in the United States of America

10 9 8 7 6 5 4 3 2

ISBN 0-13-021308-X

Prentice-Hall International (UK) Limited, *London*
Prentice-Hall of Australia Pty. Limited, *Sydney*
Prentice-Hall Canada, Inc., *Toronto*
Prentice-Hall Hispanoamericana, S.A., *Mexico*
Prentice-Hall of India Private Limited, *New Delhi*
Prentice-Hall of Japan, Inc., *Tokyo*
Simon & Schuster Asia Pte. Ltd., *Singapore*
Editora Prentice-Hall do Brasil, Ltda., *Rio de Janeiro*

CONTENTS

DEDICATION

This new edition is dedicated to my son, Nicholas. At only nine months of age, he has taught me patience, given me tremendous joy, and opened my mind to a new vision of the world. I look forward to every new day of discovery with him. I can only imagine what challenges and wonders the world will offer him when it's his time to walk.

This new edition would not have been possible without the continued love, encouragement, and support of my wife, Elizabeth. She is my inspiration for attempting to highlight the Internet as a tool for teaching and learning. I continue to be inspired by her inexhaustible energy and amazed by her skill for teaching and love for learning. Finally, I would like to thank everyone at Prentice Hall for allowing me to explore the role of technology in the classroom.

Andrew T. Stull

PREFACE
CHANGE!

I finished the first edition of this manual at 7:22 p.m. on July 23, 1995, and the second edition at 2:04 p.m. on January 31, 1997. You might think it odd that I should mention the time—but think again. At an earlier time in our history, information took months or years to cross continents and oceans. At that time, any news less than a month old was "late breaking." The accuracy of news today is obviously measured differently, now taking seconds for information to circle the globe. In fact, the popularity of the Internet has created a situation where individuals hear about important news events before the newspapers have a chance to even print them in the papers. The world has changed <u>and</u> continues to change! Our world is louder, faster, and more complex than the ones experienced by earlier generations, but it also offers more promise. In terms of information transfer, we might be described as a techno-generation; our parents, as a paper-generation. Dealing with change is a basic requirement for surviving in our modern world. But fear not, our on-line future might be chaotic <u>but</u> it should also be exciting. Prepare yourself to revel in this change.

This manual has four chapters. In the Introduction, *One Step Ahead*, I will briefly describe basic techniques and tools that you should already be familiar with. I will not, however, go into detail on many aspects that were covered in the first two editions of the *Student's Internet Guide*. There are many resources available to help beginners and novices; therefore, another beginning guide would be a waste of paper. I will gladly point you to a plethora of information to help you get started. What I hope this resource will do is help you begin using the Internet as a tool—a tool for communicating; a tool for optimizing your workload; and a tool for navigating the jungle of information available to you. Yes, you may still choose to use the Internet as a toy, and I'll even show you a few places to start, but that won't generally help you face the information challenges ahead of you.

In Chapter 1, *Finding Your Way*, you will review techniques for gleaning information from the Web. The Internet is viewed by many as a time sink and a channel for misinformation. Within this chapter you will learn about searching for, evaluating the merit of, and properly citing information. Learning to use the Internet judiciously will be a distinct asset.

In Chapter 2, *News of The Day*, you will explore ways to use the Internet to stay in touch with the news of the world and scientific happenings. The Internet, as a new media for communication, includes news sources from television, radio, newspapers, and magazines.

In Chapter 3, *Staying in Touch*, you will learn how to employ simple resources, such as e-mail, homepages, and calendars, for managing your time and organizing your busy schedule. The Internet was initially and is primarily a communications tool. Whether it be direct communication, such as with electronic mail (e-mail), or passive communication, such as with Web pages, the Internet is becoming the most prominent way for you to gather and disseminate information to the world.

In Chapter 4, *Staying in Tune,* you will learn about the tools provided for you to companion your textbook. As the digital revolution progresses, the definition of a book will change. In addition to the chapter resources found within the paper version of your textbook, there are many chapter resources found within its digital Web companion.

Throughout this guide, you will find a collection of basic and advanced resources to help you make the most of your study of science and your study of life on the Internet.

Reading this manual won't teach you all there is to know about the Internet, but it will help you to teach yourself. In the future, you will need to find information for yourself rather than rely solely on others, who may bear outdated knowledge. If you are successful, your skills in "harvesting" information from the Internet will allow you to deal with perpetual change. By the end of this manual, you should be comfortable and resourceful in navigating the complexity of the Internet, from its back eddies to its thriving thoroughfares.

INTRODUCTION
ONE STEP AHEAD

This thing that we now call the Internet has been evolving ever since it was first developed almost thirty years ago. Its prominence in our society has been increasing exponentially in recent years. It is unlikely that you are reading this manual without some basic understanding of the Internet and its features. Furthermore, I'm pretty confident that most of you have considerable knowledge of the Internet. While earlier editions of this guide were written with no expectation of your Internet experience, this is no longer a realistic position. Numerous "Internet guides" are available to help the beginner connect to and browse the Internet. This third edition of my Internet guide will take the next logical step and attempt to help you make the most out of the Internet as a tool and not simply as a toy. Let me justify this with a bit of history. At one point in time, telephones came with an instruction manual. This was a time when telephones were still new, and a telephone number was just a confusing string of numbers. Today, if you see a phone number you know what it is and how to use it. Can you imagine how funny it would be to have someone think that an explanation was needed on how to use a telephone or how to interpret a telephone number? The Internet might not be as integrated in our culture as the phone system, but it will be and not too far in the future. For a majority of you, a string of characters such as *http://www.something.com* already has meaning. Don't worry, if you're not yet this familiar with the Internet, I'm not going to leave you in the dust, but you will need to do some homework. The Web sites listed below will point you to many helpful resources about the online world. You'll be Web surfing in no time.

The Internet was born as the solution to a problem. It was designed to provide a global communication channel for the exchange of scientific information and research. Gradually, however, the Internet has also become a digital post office, a digital bulletin board, a digital telephone, and a digital tutor. Depending on whom you listen to, it may eventually be a digital television, a digital textbook, or even a digital classroom. The bottom line is that the Internet is growing in as many directions as people realize its potential and employ its power to solve their problems. But don't get too far ahead; its real merit to you is how it will solve <u>your</u> problems and make <u>your</u> day just a bit more manageable. Hopefully, that is what you'll discover here.

For those of you with a driving quest for knowledge, the resources at the following Web addresses discuss the history, growth, and culture of the Internet. In addition, I've included several basic tutorials to help you refresh your knowledge and fill in the gaps you might have about browsing, bookmarking, and connecting to the Internet.

History

History can be a valuable tool if you wish to understand the nature of things. Often, history can be used to predict the future. If you ever wanted to know why the Internet came into existence, how it has changed since its birth, or where it might go, then the following resources are the last stop you'll need to make.

BBN Timeline

http://www.bbn.com/customer_connection/timeline.htm

The BBN Timeline places important events about the Internet in context with other historical events and throws in plenty of social commentary to give you perspective.

Hobbes' Internet Timeline

http://info.isoc.org/guest/zakon/Internet/History/HIT.html

This site offers a great deal about the Internet, the people who use it, and on-line culture.

Getting Started

The following URL's are a few of the many beginner guides available on the Internet. You'll find everything you need to know about modems, browsers, e-mail, bulletin boards, chat rooms, and getting connected to the Internet on at least one of these sites.

An Introduction to the Internet from Interactive Connections

http://icactive.com/guide/index.htm

This is a very comprehensive guide to the Internet. It is provided by Interactive Connections, an Internet Presence Provider.

Learn the Net

http://www.learnthenet.com/english/index.html

Learn the Net specializes in on-line training products and services for the corporate world. Their guide is well written and up-to-date.

Net Guide from PC User Magazine

http://www.pcuser.com.au/netguide/

This guide is sponsored by the Australian version of PC User Magazine.

CHAPTER 1
FINDING YOUR WAY

Many of you reading this guide have a lot of experience with computers, while others have little or none. Before proceeding, you should have and be familiar with a few basic resources:

1. computer
2. Web browser
3. Internet connection

Don't worry if you can't afford your own. It is not my goal to spend your money. There are many free or dirt-cheap options available to you, and I'll do my best to show them to you. At an earlier time, I would not have made these assumptions, but computer labs and their use are now a common and even required component of a science curriculum. Furthermore, I'm pretty confident that these computers have one of the popular browsers by Netscape or Microsoft and an Internet connection. If you haven't found your campus computer lab yet, then my guess is that you'll find it associated with your campus library. From a beginner's point of view, the only real concern you'll have is learning the basics.

SECTION 1.1
SEARCHING THE WORLD

Although, many wire-heads consider the Internet to be the largest library on the planet, it doesn't necessarily have the easiest card catalog in the world. In this section, we'll explore techniques of searching the Internet for your gain, discuss practices for evaluating the validity of content you find, and review guidelines for citing information within your class assignments. With practice, these skills will help you improve your usage of the Internet.

There is one skill, or rather behavior, that you must adopt in order to maximize your time-to-gain ratio. That is, be aware of "search drift." The Internet is an information jungle, and if you wander into it without having a sound idea of why you are there or if you just wander around without being aware of where you are, then you will get lost and waste a great deal of time. Yes, there are times when you will want to play, wander, and have a good time, but consider whether that is important when you are trying to study the night before a test.

SEARCHING

Yahoo! is a good place to begin. It is only one of many resources available on the Internet.

I didn't name it, and I don't know how it was named, but it is easy to remember. Here is the URL for *Yahoo*!:

http://www.yahoo.com/

Yahoo! began as a simple listing of information by category—kind of like a card catalog. In recent years, it has added the ability to search for specific information. At the top level of the directory, there are several very general categories, but as you move deeper into the directory, you'll notice that the categories become more specific. To find information, you simply choose the most appropriate category at the top level and continue through each successive level until you find what you're looking for (or until you realize you're in the wrong place). Don't be afraid to experiment—it's easy to get lost but also easy to find your way home.

Suppose you were studying the impact of humans on coral reefs. Within *Yahoo*!, notice that one of the top-level categories is Science. Science seems as if would be the most appropriate place to find our information, so give it a try. After accessing the *Science* category, you'll notice that it gives a list of many different types of science. So, what is your next choice? My choice would be *Biology* but *Ecology* or even *Geology* could be possibilities; and *Marine Biology* would probably be the best. Because *Yahoo*! cross-references among the categories, you'll find that several related categories will lead you to your desired page.

Much of your success in finding information with this type of tool really centers around your preparation for the search. Often, it is possible to find information on a topic in a category that may at first seem unrelated to your topic of interest. Again, let's take the example of coral reefs and human impact. Although you may consider this a science-oriented topic, there are other avenues to consider. Aren't coral reefs a common destination for vacationers? Categories related to vacations and travel could be searched. Where are the coral reefs in the world? One of the largest is in Australia; by selecting Australia and neighboring countries you might also turn up something related to coral reefs.

Prepare yourself for a search <u>before</u> you jump into one. In the long run, it will save you both time and frustration. Don't be afraid to try some strange approaches in your search strategy. A good technique is to pull out your thesaurus and look up other names for the word. You might be able to find a more common form of the word. Think of everything associated with your question and give each of these subjects a try. You never know when you might turn up a gold mine.

The following list of resources contain many more helpful tips and techniques for searching the Internet.

Internet Search Tips and Strategies from Southern California College
http://www.sccu.edu/R-Harris/howlook.htm

Searching the Net from PC Magazine
http://www.zdnet.com/pcmag/features/websearch/_open.htm

Using Internet and Web Search Engines Effectively: An On-line Course from the American Library Association
http://www.ala.org/ICONN/advancedcourses.html

SEARCH ENGINES

A more direct approach to finding information on the Web is to use a *search engine*, which is a program that runs a search while you wait for the results. Many search engines can be found on the Web. Some of Web search engines are commercial and may charge you a fee to run a search. Search engines are also available for other parts of the Internet: *Archie, Veronica*, and *Jughead* are examples of such search engines.

As mentioned earlier, *Yahoo*! also has a useful search engine. A search engine that I use frequently is called *Lycos* (*http://www.lycos.com*). It's simple to operate but, as with any search tool, it takes practice and patience to master. Take the time now to connect to *Lycos*, and we'll take it for a test run. When you first see the opening page, you'll notice that it is very complex. But it's an excellent resource, and the instructions on the page will tell you almost everything you need to know. To search, enter a word into the white text entry box and press the submit button. *Lycos* will refer back to its database of information and return a page of hyperlinked resources containing the word you entered. When the query results come back to you, notice that they are hyperlinks to various sites on the Internet.

To see how a search engine works, use *coral reef* as a topic for a search. Notice that you can set the number of responses that the engine will return to you. Did you notice that some of your results didn't seem to apply to your topic? This is one of the pitfalls of search engines. They are very fast, but they don't think—that is your job. A search using the term *coral reef* is just as likely to turn up a link to Jimmy Buffett's Coral Reefer Band as a link to coral reef research. To perform an effective search, you will need to spend time <u>before</u> the search preparing a search strategy. When you do research using an automated tool like a search engine, you can expect many links to be unrelated to your topic of interest—but all in all, search engines are still very powerful tools.

Another type of search service that you'll hear much about is called a meta-search service. This type of service will send your query out to a number of different search engines and then tabulate the results for you. They come in many different levels of sophistication and they also generate a large amount of information. If you're not intimidated by volume then give one of them a try.

> Here's a meta-search tool that I just learned about and one that is actually fun to use. Give it a try.
>
> **Ask Jeeves**   **http://www.askjeeves.com/**

One last word on search engines. These tools don't directly search the Internet. They actually search a database that is derived from the Internet. Here is how it works. Search engines use robots (automated programming tools) that search for and categorize information. This information is placed into a database. It is this database that you search when you use the search engine. Can you think of a potential problem with this system? Unfortunately, the quality of the database depends on the effectiveness of the robot that assembles the database. This is why you should not rely on just one search engine tool. Use several, because what one does not find, another might.

> The following resources will help you learn more about searching the Internet.
>
> **How To Search the Web from Palomar College**
> http://daphne.palomar.edu/TGSEARCH/
>
> **Search Engine Tutorials from Mercklermedia**
> http://searchenginewatch.com/resources/tutorials.html
>
> **Search Engine Watch from Mercklermedia**
> http://searchenginewatch.com/
>
> **World Wide Web Research Tools from Southern California College**
> http://www.sccu.edu/faculty/R-Harris/search.html

SECTION 1.2
BEING A LITTLE PICKY

In your career as a student and eventually as a professional, you will spend a great deal of time using the Internet to communicate and find information. But, can you trust the information you find? In traditional publications, just as with Internet publications, there are strong, reliable sources of information and then there is the other end of the spectrum. Is it possible to leave a grocery store without passing a tabloid newspaper displaying a title like "Elvis Was A Spy for a Race of Alien Invaders?" It's obvious that this title is misleading as we all know that Elvis was actually a double agent and on our side. When information is outlandish, it is easy to spot the truth from the lie, but not everything is as obvious. Misinformation is occasionally passed on by respected publications as well. Do you remember CNN's 1998 retraction of an erroneous report saying that US soldiers used the nerve gas sarin in a Vietnam-era mission? The truth was uncovered but not before it was passed to the world. Yes, the system is often self-correcting, but this takes time. Many people in our society view "first and incorrect" as being more important than "second and correct." With the speed and ease with which we can "publish" information on the Internet, there is little time for the system to identify its own mistakes. Developing skills to evaluate all sources of information intelligently, especially those from the Internet, will be a valuable asset.

Evaluating the merit and accuracy of an information source isn't new to the Internet. Criteria for evaluating content has been around ever since it was possible to pass infor- mation from one person to another. The informal criteria that have evolved through time have simply been adapted to reflect each advance in our ability to communicate. Criteria for evaluating the Internet have been adapted from the previous standards to reflect the unique challenges offered by the speed and global nature of the Internet.

The following is a list of criteria that I use to evaluate information sources. They are adapted from traditional evaluation criteria and personal observations of the behavior of Internet authors, managers, and publishers. Additionally, there are many Internet sites that discuss evaluating information from the Internet. You may eventually develop additional criteria by which you evaluate Internet sources, but these should get you started:

1. Authority
   Who is the author?
   What are the author's credentials or affiliations?
   Is contact information listed for the author?
   Does the source cite respectable references?

2. Accuracy
   Does the piece follow basic spelling, grammar, and composition rules?
   Does the information appear to be reliable?
   Are the embedded hyperlinks valid?
   Do the hyperlinked or referenced sources present accurate information?
   Does the author accept feedback and error notices?

3. Objectivity
   How is the source biased?
   Does the author have an agenda?
   What is the purpose or intent of the piece?
   Is the piece intended to be persuasive?

4. Currency

When was the original piece created?

Is the piece updated on a regular basis?

Does the piece incorporate recent events and information?

5. Coverage

Who is the intended audience?

How comprehensive is the piece?

Does the author present the depth as well as the breadth of the topic?

Is the content of the piece original, an interpretation, or a reference work?

6. Stability

Does a student, an instructor, or an institution author the piece?

Is the URL likely to change over time?

What is the domain and URL of the posted piece?

What is the primary focus of the organization behind the sponsoring domain?

Is the piece part of a managed and maintained Web site?

Does the site have an identifiable site administrator?

How long has the piece been resident at the current URL?

7. Utility

Is the piece valuable as a source of primary information for a topic?

Is the piece valuable as a source of reference information for a topic?

Does the piece effectively use visual media and graphic enhancements?

Is the piece a reference tool such as a bibliography or link library?

Is the piece a service tool such as a dictionary, glossary, or search engine?

Is the piece a communication tools such as a newsgroup or chat room?

Of course, my list of criteria is not the last word on the topic. If you wish to review criteria that others have proposed for evaluating Internet resources, look through these sites. Each will offer you a unique point of view but all are valuable sources of information that I encourage you to read.

Evaluating Quality of the Net from Babson College
http://www.tiac.net/users/hope/findqual.html

Evaluating Internet Resources: A checklist from University of California, Berkeley
http://infopeople.berkeley.edu:8000/bkmk/select.html

Evaluating World Wide Web Information from The Libraries of Purdue University
http://thorplus.lib.purdue.edu/library_info/instruction/gs175/3gs175/evaluation.html

Ten C's for Evaluating Internet Resources from University of Wisconsin-Eau Claire
http://www.uwec.edu/Admin/Library/10cs.html

SECTION 1.3
LISTING WHAT YOU FOUND

The next logical step after you've found and evaluated your sources of information is to correctly cite them in your work. Just as everything else in science, citing references requires a specific format. A natural part of all scientific reports is a citation list supporting the background, design, and conclusions of the described research. Different scientific disciplines and journals will have different formats but in general there is a common design by which scientific references are cited. For example, the Council of Biological Editors has a standard style for cited references. As we've seen with the other topics included in this chapter, the traditional way of doing something often needs to be modified to include on-line content. It is a simple matter of properly citing on-line references so I will not go into it here. A number of print and on-line reference works are available to help you. Additionally, many journals will have their own format for listing on-line references. To reiterate what I wrote in the section on evaluating content, it is important to properly evaluate your references so that they have merit as a cited work.

CHAPTER 2
NEWS OF THE DAY

The Internet is proving itself to be a fast means of distributing news to the world. Even before the news agencies have a chance to convert news into ink and paper, it is available to you on the Internet. Your problem now is not one of news <u>access</u> but one of news <u>volume</u>. Yes, you may still be asking <u>why</u> you should be concerned with science happenings. The simple answer to this question concerns power, control, and security. If you want to be able to decide how the world around you changes, what medical options are best for you, and when pending legislation is bad for you and your family, then knowledge of science is a must; and the fastest track to awareness of scientific events is through the news. For the pragmatic among you, it is also a great source of information for reports and current event assignments.

In this section of the chapter, we'll explore a few of the many sources of general and science news on the Internet, and you'll see how easy it can be to surf the Internet without being buried by the information wave.

SECTION 2.1
A SMORGASBORD

Science is an everyday tool by which our society addresses and challenges our modern age. With the prominence of the Internet, it is very simple to stay on top of the happenings and events in science so that you can influence your life's course.

In the past, each news media had a distinct format by which it distributed its product. Although distinct in the past, these traditional news media have each tapped into the Internet. The lines that were once distinct between newspapers and television are now blurring together in the Internet.

The following list of organizations offers convenient access to science news. The URL listed for each is the general Web address for the news source. Upon reaching this page, you can choose a number of different categories of news. Since it says Science on the cover, we should probably talk about that. In a majority of the cases, "Environment," "Health," "Science," "Technology," or "Sci/Tech" are categories that will lead you to news with a science focus. Most of these are traditional media but have added an Internet format to their service so you are likely to find a similar article in their traditional media deliveries, newspaper, radio, and television. Often, if you see or hear of a newspaper, radio, or television piece, you can go to their Internet page for details and related resources.

Television	**Web Address**
ABC News*	http://www.abcnews.com/
BBC News	http://news.bbc.co.uk/
CBS News*	http://www.cbs.com/navbar/news.html
CNN	http://www.cnn.com/
Discovery On-line	http://www.discovery.com/
FOX News	http://www.foxnews.com/
MSNBC*	http://www.msnbc.com/news/
NBC News*	http://www.nbc.com/nbc/news_pg.nbc
PBS	http://www.pbs.org/

Newspapers/Radio	**Web Address**
BBC World Service	http://www.bbc.co.uk/worldservice/
LA Times	http://www.latimes.com
NPR (National Public Radio)	http://www.npr.org/news/
NY Times[#]	http://www.nytimes.com
Science-Friday	http://www.sciencefriday.com/
Seattle Times	http://www.seattletimes.com
USA Today	http://www.usatoday.com

In a few cases, those marked by the *, news will be customized to reflect the events in your local area if you return their request for your zip code. Occasionally, some of these news services may require you to register to receive their news but most are entirely free. The common model for these Internet news sources is similar to that of commercial television. Yes, you've got to wait through the soap commercials in order to watch your shows about murder, mayhem, and sex. Alternatively, if you want your news without commercials and you don't have the time to peruse the Internet, there is a resource you may want to consider. College NewsLink (http://www.ssnewslink.com/) is a service, affiliated with the publisher of this guide. For your subscription fee, Newslink will gather, organize, and deliver topic-specific news to you through your e-mail every day. These articles are selected from a large list of newspapers from around the world. Furthermore, each article is embedded with links to related sources. Although, it is not a free service, it might offer significant time advantages for you.

SECTION 2.2
BY THE MENU

Although television and newspaper Internet sites may offer interesting and informative articles, they probably don't offer the detail or emphasis that is at the heart of a scientific publication. For this reason, you may find that you need a science-related source of information without the politics and social commentaries. The following list of sites are an excellent place to begin your search for this type of information. Many are free but a few require you to subscribe to their service.

<u>Science Information</u>	**<u>Web Address</u>**
American Scientist*	http://www.sigmaxi.org/amsci/amsci.html
exoScience	http://exosci.com/
EurekAlert	http://www.eurekalert.org/
InScight	http://www.academicpress.com/inscight/
NASA	http://www.nasa.gov/
Nature*	http://www.nature.com/
Newswise	http://www.ari.net/newswise/
Newslink*	http://www.ssnewslink.com/
Quadnet	http://www.quad-net.com/
Science*	http://www.science.com/
ScienceDaily News	http://www.sciencedaily.com/index.html
Science News*	http://www.sciencenews.org/
Scientific American*	http://www.sciam.com/
Unisci	http://unisci.com/

A few of these are peer-reviewed research journals for professional scientists; therefore, potentially too expensive for you. As I said, it's not my hobby to spend your money. If you need an article from one of these on-line journals, then your library may have a site subscription or one of your professors may have a research subscription. Don't open your wallet before you try all your options.

The last thing you should remember about on-line news services is that most of them offer a search and archiving tool. Often, their articles are aggressively hyperlinked to related materials throughout the Internet. Although you won't see something of interest at the top of the page, you might find something interesting with a quick search. Further, an article of average interest might take you to something truly valuable with one of their embedded links.

CHAPTER 3
STAYING IN TOUCH

Although the Internet is commonly thought of as a flashy, graphically rich waste of time, it began as a tool to enable researchers to communicate between research labs across the United States. I hope your mind is open or that I can dissuade you from this jaded view. If you look at its basic features, the Internet is still a valuable and effective tool for communication. As you've learned in Chapter 1, this technology is helping to make a big world small. In essence, one goal of the Internet has been, and hopefully will continue to be, to eliminate the hindrance of geography on the free exchange of ideas. Whether it becomes a waste of time or a timesaving tool is entirely up to you. I hope the following ideas will help you make the most of the Internet as a tool for communication and collaboration.

SECTION 3.1
A MAILBOX IN CYBERSPACE

An e-mail account is the most basic of methods for planting yourself in the Internet community. I am continually surprised with how pervasive e-mail has become. My 90-year-old grandmother even has an e-mail account! Do you? Don't worry if you don't, I have a number of simple, cheap, fast solutions you may want to consider.

In case you still need encouragement, here are a few reasons to help persuade you. An e-mail account gives you an identity and a point of contact on the Internet. It's cheaper than the phone and it's faster than postal mail. In a world that is progressively more connected, not having an e-mail address is paramount to being invisible. I'm not saying that you will cease to exist, but I am saying that when more and more people view e-mail as a basic form of communication, you will be at a distinct disadvantage. When all of your classmates are collaborating on a group assignment, helping each other to learn a difficult concept, or planning to gather for a crash study session or a pizza, it might not be easy for you to stay in the loop. In addition, many of your instructors have begun using e-mail to communicate with students. You can probably come up with a list of disadvantages, but I think you'll find that there is an overall advantage for you to have an e-mail account.

Now that I've convinced you that an e-mail account is in your future, there are a few options available to you. Most of you can apply for an e-mail account through your college. If your college doesn't provide student e-mail accounts, then e-mail service through an Internet Service Provider (ISP) is a second option. ISP's require you to subscribe (meaning spend money) to acquire their service. The nature of service, hourly or monthly, will depend on your anticipated use. The disadvantage is that you will need to pay a fee for the service. This can also be considered an advantage because you can expect help from time to time, which you are not as likely to receive from other options.

A third option, increasing in popularity, is to choose a free e-mail service provided by one of the many on-line companies. Yes, a free e-mail account with many of the bells and whistles found in regular e-mail account can be yours for the asking. If you choose a free e-mail service, then read the fine print and understand what it means to you. In most cases, the service is provided to you free because the provider is making its money by selling advertising space to others companies. This is the same thing as search engine companies do. In order to read your mail, you have to wade through a few commercials prominently posted on your e-mail reader. An additional condition of these free e-mail accounts is that the service will gather information about you in order to customize and target the display of commercials for you. In most cases, this information is used only to target you with commercials but, as I've said already, read the fine print.

> The following are only a few of the more prominent services offering free e-mail. Read the fine print in their service agreements, and choose the one that offers you the most.
>
> | **E-Mail Service** | **Web Address** |
> | --- | --- |
> | Hotmail | http://www.hotmail.com/ |
> | Netscape | http://webmail.netscape.com/ |
> | Yahoo! | http://mail.yahoo.com/ |
>
> The following list of on-line resources should help you find, understand, and use your e-mail account.
>
> **A Beginner's Guide to Effective Email by Kaitlin Duck Sherwood**
> http://www.webfoot.com/advice/email.top.html
>
> **ICONnect On-line Courses on E-Mail from the American Library Association**
> http://www.ala.org/ICONN/ibasics2.html

SECTION 3.2
A PLACE TO CALL HOME

After an e-mail account, a home page is the next logical step toward establishing yourself with an Internet presence. Considering the proliferation of personal home pages, and the typical merit of their content, you might not realize the advantages that a personal home page may offer to you. While an e-mail account offers you an identity on the Internet, a home page offers you a central resource that is mostly under your control. As a student, you are somewhat nomadic and therefore required to work in many different locations through out the day. A home page can be an important central resource for your nomadic life. Your home page could list on-line reference sites such as search engines, dictionaries, directories, and glossaries; a hyperlinked list of e-mail addresses for your instructors, classmates, and friends; a place where you can post shared information for your study groups; or a place to post class assignments for your instructors. In short, a home page may be passive in nature, but it can be a valuable tool for communication.

You have three basic options for posting and maintaining a home page on the Internet. Your college may offer you space to post and maintain a home page, you can subscribe to an ISP, or you can use a free service. The business model used by free e-mail services is similar to that of companies providing free home page services. In most cases, these services have a basic format that you can modify and occasionally add to. Read and understand the fine print to know what you are agreeing to.

The following services enable you to set up a home page on the Internet. Each of them offers a slightly different service, so spend a bit of your time to really evaluate their offerings.

Home Page Service	**Web Address**
Geocities	http://www.geocities.com/join/
Microsoft	http://home.microsoft.com/
Netscape	http://my.netscape.com/
Yahoo!	http://my.yahoo.com/

The following list of on-line resources should help you begin building your first home page. With a quick search of the Internet, you will find a large number of other resources along this line. Be creative and enjoy the experience.

A Beginner's Guide to HTML
http://www.ncsa.uiuc.edu/General/Internet/WWW/HTMLPrimerAll.html

The Bare Bones Guide to HTML by Kevin Werbach
http://werbach.com/barebones/

Writing HTML from Maricopa Community College
http://www.mcli.dist.maricopa.edu/tut/

SECTION 3.3
A CALENDAR OF EVENTS

The final step in my project to help you stay in touch with your commitments to your classes, friends, and family is to make you aware of the help that a calendar program can lend. By now, it should be obvious to you that your life is not going to get less complicated. Having a tool to help you schedule your time and remember important events will be a distinct asset. Developing a routine to organize your life is the first and best step to take. The second step is to find a tool to help you remember your reading and homework assignments, library time, class schedules, exams, study group meetings, office hours for your instructors, in addition to all of your personal commitments.

As always, your options are numerous. Memory is probably the most common option initially employed by the novice, but I think you'll quickly determine its disadvantages. There are also paper calendars specifically designed to help you organize your time. A number of companies have specialized in paper-calendar organizing systems. Although they may have many successful customers, you must consider whether a calendar book is for you. The next step in the progression is calendar software programs. These programs have the obvious advantage provided by computers—most prominently, the ability to automatically track and remind you of important events. The most obvious disadvantage for such a program relates back to your nomadic life style. Can you really get to your calendar program when and where you need it? Lest you think that you are out of options, let me lead you to yet another of those free on-line resources. It is not necessarily your best option, but it has advantages. In addition to offering you free e-mail, Yahoo! will allow you to use their free calendar service, provided you register. By now you should be familiar with the model. The service is free to you, but you'll need to provide them with some basic personal information and you will need to endure the targeted commercials embedded in your calendar viewer. Once again, read and understand the user agreement before you accept their offer. With their service, you will be able to populate a calendar with events that are important to you. Your calendar of events is viewable by day, week, month, or year. It will contain both a "To Do" list and a regular daily schedule. One of the potentially valuable resources is that of scheduled e-mail notes to remind you of important events. Never again do you need to suffer those nightmares of forgetting an exam. However, you do need to make the commitment to maintain the accuracy of your calendar. Additionally, if you know basic HTML, you can schedule events to include hyperlinks. These could be to reference sites, assignments posted by your instructors, or to resources posted on the Companion Web site for your textbook. Essentially, your calendar can be completely linked to the Internet.

I have not described other free on-line calendar tools because I have not as yet encountered them. Fear not, by now you should have the skill and savvy to find other comparable services. If Yahoo! is indeed the first to provide this service, then I suspect that others will soon pop-up with new and better features. Recently, competition has had a strange way of influencing the evolution of the Internet.

Parting advice, remember that you have the power of the purse; therefore, always look for the cheapest option, read and understand an agreement before you sign it, and enjoy your journey.

SOMETHING CALLED PRIVACY

The Internet has been moving over the last few years to increase the level of personalization that users experience. You may have seen this on many of the news sites providing customized news delivery to your area. It only requires that you enter your zip code. Yes, it saves time and seems friendly to have a page that automatically loads with weather and happenings for my home town but it also worries me that someone, or something, might be recording more about me than I like.

There are two aspects to personalization that you should be aware of. In some cases, information about you is not actually sent to the party requesting the information. Instead, it is stored on the machine you are using to access the Internet; therefore, it is also available to them each time you come back to their site. For example, if you did give your zip code to one of the news companies I described earlier, that zip code information is stored on your machine as something called a cookie. It is called to action each time you log onto the news site and is used to customize the page with information that is appropriate to your area. If by accident or intention, you entered a zip code for someplace 2000 miles away, then you can probably predict the outcome. It's a simple matter to erase the zip code information, but it requires an extra step on your part.

The previous example was only one type of problem you might experience with cookies. Consider the situation where you and several other people, possibly room mates or lab mates, all share the same computer. It is likely that they will configure cookies that reflect their preferences and not yours. This could be their e-mail address, home page address, the size of their shoe, their age, or a number of other things. This is obviously a problem that you'll experience in a campus computer lab. In many computer lab situations, the administrator will disable the cookie feature in the browsers and deny everyone access to customizing features. If by chance you do share a computer with many different people and the cookie feature is enabled, you must be aware of the problem whenever you make a request.

A second aspect of privacy and the Internet deals with subscription information. This information, unlike information that resides on your local computer, is often sent to the requesting server. In some cases, a cookie is sent to your local machine to identify you, and in other cases, you are requested to log in each time you come back to the subscribed service. Both situations can cause you difficulties if you are unaware of how your information is being handled. As I've said a number of times, before you send anything about yourself through a browser, you should read the fine print and know what you are requesting and what is happening to your information.

There are also plenty of powerful people worried about privacy on the Internet, so we are not alone. The following Web address are for two of the many organizations that have dedicated themselves to securing and lobbying for Internet privacy. It might be a helpful exercise to make a visit to their site and learn more about the situation.

Electronic Privacy Information Center
   http://epic.org/

Electronic Frontier Foundation
   http://www.eff.org/

I'm not trying to create a sense of mistrust in you about using the Internet. I don't want you to confuse the Internet with the X-Files. The great majority of times that you are asked to supply information on the Internet, it is safe to do so and is meant to help the requester more fully service your needs. But I am saying that you must be aware of conflicts that can happen when you share resources such as a computer.

CHAPTER 4

STAYING IN TUNE

Change is a central theme of science and this guide. As you progress through college I think that you will continue to realize that change is also a significant aspect to all material things. You should begin to notice that textbooks are not immune to this phenomenon either. Not that long ago, books were tremendously expensive because they were individually hand written and hand illustrated. As such, they were great works of art owned by the elite but very poor resources for reaching the masses. As is typical of our species, problems demand solutions, and the printing press was invented—probably the most significant bit of technology to have appeared in the last millennium, enabling the mass production of affordable books.

In its basic form, a book is a collection of words and images organized to communicate one or more ideas. As alternatives to books, song, dance, and plays are also methods used to organize both words and images for communicating ideas. I'm trying to make the point that paper and ink are only artifacts and not necessarily integral to the process of communication. The Internet has presented us with the next significant step toward communication and ubiquitous access to information. This last chapter will describe a few of the textbook-specific Internet resources that are available to you. Hopefully, you will see a glimmer of the future of information, education, and books through your experience with these resources.

SECTION 4.1
A WEB COMPANION

A picture says a thousand words. I don't know who first made this statement, but not many of you will doubt its truth. Often, an image more easily communicates an idea than does a written or spoken description. Now, if a still image can say so much, how much can a video or an animation say? I suspect that each can communicate a great deal if they are presented in the correct way. Written and spoken descriptions will always have a major place in education because visual resources have their limitations too. Often, the specificity of a word cannot be replaced by the general impression of a picture. Forming the proper combination of words, images, animation, and even sounds will be the challenge of textbook authors in the future. Toward this end, your textbook now includes an added digital tool chest, an alter-ego if you will, called a Companion Web (CW) site. As a resource it employs the unique character of the Internet and contains many tools to help you visualize, communicate, and discover concepts introduced in your paper textbook. In the future, I suspect that we will see the definition of book expand to include digital content as well as the traditional content more conveniently delivered with ink and paper.

The next obvious question is "How do I find the CW site that goes with my book?" Fortunately, there is an easy answer to this question. All Prentice Hall textbooks have a convention for addressing their CW sites. The last name of the first author of the textbook is used to distinguish one site from another. For example, if you are in an Ecology class, you are probably using *Ecology: Theories and Applications* by Peter Stiling. If you add Stiling's name to the standard Prentice Hall Web address (http://www.prenhall.com/) then you will find the CW for his book.

http://www.prenhall.com/stiling/

In this fashion, you should be able to find a CW for any book if you know the author. Because you might not know the author for a book on every subject you'll want, there is an indirect way to reach all of the CW sites. Simply load the Prentice Hall home page (http://www.prenhall.com/) and select the Companion Website Gallery option from the page. All of the Prentice Hall CW sites are organized by discipline and are available from the CW Gallery. Here is a partial list of science CW addresses to give you a perspective of the breadth of resources available to you.

Science Topic	**Web Address**
General Biology	**http://www.prenhall.com/audesirk/**
Astronomy	**http://www.prenhall.com/chaisson/**
Physics	**http://www.prenhall.com/giancoli/**
Genetics	**http://www.prenhall.com/klug/**
Anatomy and Physiology	**http://www.prenhall.com/martini/**
Environmental Science	**http://www.prenhall.com/nebel/**
Earth Science	**http://www.prenhall.com/tarbuck/**

Although I've only listed science CW addresses, there are CW sites that support every major discipline that you'll encounter in college. In addition, if your professor is not using a book with a CW site then you can always browse through the CW Gallery until you find a resource to help you with any topic.

SECTION 4.2
A WEB TOOL BOX

Now that you know where you can find your CW, here's what you'll see when you look inside. Just as a tool box is a container for tools, so too your CW is a container for unique Web tools. In addition to general tools, you are likely to find a few discipline-specific tools. Each CW site was assembled to match the unique flavor of the textbook's discipline, and so you might find a molecular model viewer in a chemistry site but not in a geology site.

All CW's share a few basic tools accessible from the first page of their site. The CW site for Peter Stiling's Ecology textbook is a good example. In addition to an image of the textbook, you will find the following three features.

SYLLABUS MANAGER

This tool both allows your professor to create and manage an on-line syllabus for your course <u>and</u> enables you to view her syllabus as part of your CW. This tool is valuable to you only if your professor first develops an on-line syllabus with it. If the professor hasn't but you think that it would be helpful, then you might consider offering to help. You can learn more about this tool by taking the *Syllabus Creation Walk-Thru* found on the Syllabus Manager page.

An on-line syllabus will enable you to reach assigned activities within the CW site, sites for assignments outside the CW, your professor's individual Web pages, in addition to the many on-line tools available to you on the Web.

If your professor already has an on-line syllabus in the Syllabus Manager tool then you only need to find it to use it. To do this, simply use the search selector on the Syllabus Manager page, type in your professor's name or your college, and select the "Search Now" button. A list of on-line syllabi will be displayed from which you can select your course. Once you select your course, a calendar will appear in the lower left corner of the browser page. This calendar will contain dates and assignments as posted by your professor. It's actually pretty easy to use, but if you need more help you can select the Help tool also found on the front page of every CW site. I'll explain more about this later.

There is another scenario that you should be aware of. If you are using a computer in a campus lab, you may occasionally find that other students will leave their syllabus loaded when you sit down to use the computer. The last option at the bottom of the syllabus calendar will unload the last syllabus and return you the menu where you can select your own syllabus. Additionally, even if you sit down to a computer without a loaded syllabus, within the syllabus request window, you may see a list of syllabi for other courses and professors. If you've already used this tool, you might even find that your course is included in this list. You have the option of selecting your course from the list or searching for it again. Give it a try and you'll discover that my description of the process is actually more complex than the actual process itself.

YOUR PROFILE

If you haven't done so, please read the section in Chapter 3 that deals with your privacy. This tool is intended to help make your use of the CW resources more rewarding. The information that you enter into this tool is used to help you customize your experience with the CW site. Once this information is entered, when you come back to this site, some of the basic features will be preset and ready for your use. As I mentioned earlier, be aware of the situation where someone else might use information you preset or where you might use information they preset in a shared computer. If you do share a computer then you should probably skip over this tool.

HELP

It's all in the name. If you are having a difficult time using one of the chapter tools, configuring your browser, or using the Syllabus Manager then you can find help here. Probably the most helpful of the resources you'll find in Help is called the Browser Tune-Up. It is a resource that will diagnose your browser and its plug-ins to determine if you have the latest versions of the software. If you do not have the latest versions, you can download them through the tool and test them to make sure they are working properly. It will probably be helpful to do this on a periodic basis in the event a new version of the software is released.

SELECT A CHAPTER

The chapter selector is the doorway to the meat of the CW site. All of the tools described up to this point were general and present in all CW sites, but the resources you'll find within each chapter of the CW will match the topic and intent of the chapters in your textbook. This aspect of the CW is organized in parallel to the textbook. Notice that you can navigate anywhere in the CW with the navigation bar on the left side of the CW window. This vertical bar also lists all of the tools that are available to you within the chapter. As you may have discovered in reviewing the Help tool, all aspects of the CW are thoroughly described there. I encourage you to review the different tool descriptions and to play around with each in the chapter area of your CW. Each of the tools that you'll encounter within a chapter was designed to help you understand the topics presented in the chapter.

Although I will rely on your curiosity to discover the character of each tool, I would like to make a special note of the Feedback feature. The Internet is a dynamic place. It is not just different every day but also every second of every day. If you find something that you think will be helpful to others taking this course, that you feel is a mistake, or that could improve any of the CW sites, please send in feedback and the site will improve for everyone.

IT'S ALL GREEK TO ME

Access Provider

This is a company that provides access to the Internet or a private network for a fee. (See Internet Service Provider.)

Agent

This is a type of software program that can be directed to automatically search the Internet or perform a specific function on behalf of a user. Spiders and worms, which roam the Internet, are the most common types of agents.

Anchor

This is an HTML tag used by a Web page author to designate a connection between a word in the text and a link to another page. (See HTML, Tag, and Link.)

AVI

This stands for Audio/Video Interleaved. It is a Microsoft Corporation format for encoding video and audio for digital transmission.

Backbone

This is the main network cable or link in a large internet.

Bandwidth

This term refers to the capacity of a network line to carry user requests. Network lines such as a T1 are larger (have a higher bandwidth) and can carry more information than a lower bandwidth line such as an ISDN or a modem connection. (See ISDN, and Modem.)

Bookmark

This refers to a list of URLs saved within a browser. The user can edit and modify the bookmark list to add and delete URLs as the user's interests change. Bookmark is a term used by Netscape to refer to the user's list of URLs; Hotlist is used by Mosaic for the same purpose. (See Hotlist and URL.)

Browser

This is a software program that is used to view and browse information on the Internet. Browsers are also referred to as clients. (See Client.)

Bulletin Board Service (BBS)

This is an electronic bulletin board. It is sometimes referred to as a BBS. Information on a BBS is posted to a computer where people can access, read, and comment on it. A BBS may or may not be connected to the Internet. Some are accessible by modem dial-in only.

Cache

This refers to a section of memory that is set aside to store information that is commonly used by the computer or by an active piece of software. Most browsers will create a cache for commonly accessed images. An example might be the images that are common to the user's home page. Retrieving images from the cache is much quicker than downloading the images from the original source each time they are required.

Chat room

This is a site that allows for real-time person-to-person interactions.

Client

This is a software program used to view information from remote computers. Clients function in a Client-Server information exchange model. This term may also be loosely applied to the computer that is used to request information from the server. (See Server.)

Computer Virus

This is a program designed to infect a computer and possibly cause problems with the infected system. Viruses are typically passed from user to user through the exchange of an infected file. Numerous virus checkers or scanners are available to help you identify and inoculate your system against viruses.

Compressed file

This refers to a file or document that has been compacted to save memory space so that it can be easily and quickly transferred through the Internet.

Cookie

A cookie is a small piece of information given temporarily to your Web browser by a Web server. The cookie is used to record information about you or your browsing behavior for later use by the server. For example, when you visit an on-line bookstore, a cookie will probably be passed to you browser to record book selections you make for purchase.

Cyberspace

This refers to the "world" of computers. It was coined by William Gibson in the novel Neuromancer.

Dial-Up Account

This refers to having registered permission to access a remote computer by which you are allowed connect through a modem.

Domain

This refers to the different subsets of the Internet. The suffix found on the host name of an Internet server defines its domain. For example, the host name for Prentice Hall, the publisher of this book, is www.prenhall.com. The last part, .COM, indicates that Prentice Hall is a part of the commercial domain. Other domains include .MIL for military, .EDU for education, .ORG for non-profit organizations, .GOV for government organizations, and many more.

Download

This is the process of transferring a file, document, or program from a remote computer to a local computer. (See Upload.)

E-mail

This is the short name for electronic mail. E-mail is sent electronically from one person to another. Some companies have e-mail systems that are not part of the Internet. E-mail can be sent to one person or to many different people. (I sometimes refer to this as Junk E-mail.)

Encryption

This refers to a security procedure of coding information to prevent unwanted viewing. Information sent across a computer network is typically disassembled, shipped, and reassembled on the receiving computer. Encrypted information must be decrypted with a special "encryption key" by the receiving party.

Executable File

A file or program that can run (execute) by itself and that does not require another program is said to be an executable file. Some files, such as word processor documents, require an applications program for viewing them.

External Viewer Application

Browsers are software applications that enable users to display content distributed on the Web. Web information must be in one of a few specific formats before the browser can display it for the users. An External Viewer Application can be used to view files sent across the Web that cannot be viewed within the browser. These applications are said to be external because they do not operate within the browser. (See Plug-in.)

FAQ

This stands for frequently asked questions. A FAQ is a file or document where a moderator or administrator will post commonly asked questions and their answers. Although it is very easy to communicate across the Internet, if you have a question, you should check for the answer in a FAQ first.

Firewall

A firewall is a network server that functions to control traffic flow between two separate networks. They are typically used to separate large government and corporate sites from the Internet. Some colleges use firewalls to protect certain ares of their network.

Flame

Degrading a person over the Internet is referred to as flaming. Non-verbal communication is not typically possible across a computer network, unless you have a video hookup, so misunderstandings often result. Anonymity of the flamer also contributes to such an exchange because people are more likely to make impolite statements given their separation.

FTP

This stands for File Transfer Protocol. It is a procedure used to transfer large files and programs from one computer to another. Access to the computer to transfer files may or may not require a password. Some FTP servers are set up to allow public access by anonymous log-on. This process is referred to as Anonymous FTP.

GIF

This stands for Graphics Interchange Format. It is a format created by CompuServe to allow electronic transfer of digital images. GIF files are a commonly used format and can be viewed by both Mac and Windows users.

Gopher

This is a format structure and resource for providing information on the Internet. It was created at the University of Minnesota.

GUI

An acronym for Graphical User Interface. Macintosh and Windows operating systems are examples of typical GUIs.

Helper

This is software that is used to help a browser view information formats that it could not normally view.

Hits

This refers to a download request made by a browser to a server. Each file from a Web site that is requested by the browser is referred to as a hit. A Web page may be composed of numerous file elements and although hit counts are often reported as a measure of popularity, they are misleading.

Home Page

In its specific sense, this refers to a Web document that a browser loads as its central navigational point to browse the Internet. It may also be used to refer to as Web page describing an individual. In the most general sense, it is used to refer to any Web document.

Host

This is another name for a server computer. (See Server.)

Hotlist

This is a list of URLs saved within the Mosaic Web browser. This same list is referred to as a Bookmark within the Netscape Web browser.

HTML

This is an abbreviation for HyperText Markup Language, the common language used to write documents that appear on the World Wide Web.

HTTP

An abbreviation for HyperText Transport Protocol, the common protocol used to communicate between World Wide Web servers.

Hypertext

This refers to an embedded connection within a Web page that connects to a place within the viewed Web page or to a different Web page. Web pages use hypertext links to call documents, images, sounds, and video files. The term hyperlink is a general term that applies to elements on Web pages other than text.

Icon

This refers to a visual representation of a file or program as it is represented on a typical windows graphic user interface (GUI). For example, Apple uses a trashcan icon to represent the place where you put files you want to delete or remove from your computer. Microsoft uses a wastepaper basket.

Internet Relay Chat (IRC)

IRC is a network attached to the Internet. It allows users to converse in real time with other individuals. It is not typically a one-on-one conversation. Chat "rooms" are typically a very confusing place for beginners.

Internet Service Provider (ISP)

A company that provides Internet access is an ISP. Your ISP might be your school or a company to which you subscribe on a monthly basis.

Intranet

This refers to a network of networks that does not have a connection to THE Internet.

ISDN

This stands for Integrated Services Digital Network. It is a digital phone line. ISDN service is typically more expensive but also offers customers added features such as a greater bandwidth. (See Bandwidth.)

Java

This is an object-oriented programming language developed by Sun Microsystems.

JavaScript

This is a scripting language developed by Netscape in cooperation with Sun Microsystems to add functionality to the basic Web page. It is not as powerful as Java and works primarily from the client side.

JPEG

This stands for Joint Photographic Experts Group. It is one of the common standards for pictures on the Internet.

Local Area Network (LAN)

A LAN is a small or local network, typically within a single building.

Link

This refers to a text element or graphic within a document that has an embedded connection to another item. Web pages use links to access documents, images, sounds, and video files from the Internet, other documents on the local Web server, or other content on the Web page. Hyperlink is another name for link.

List Administrator

This is an individual that monitors or oversees a mailing list. (See Mailing List.)

Login

Generally, this refers to the act of connecting to a network but it may also indicate the need to enter a username or password to access a network or server.

Lurker

This is an individual that connects to a chat room, bulletin board, or newsgroup and observes the conversation but does not participate.

Mailing List

This is a functional group of e-mail addresses intended for making group mailings. It is used as a simple bulletin board. Some mailing lists are moderated by an individual and some are automatic. The most common mistake made by people using mailing lists is that they reply to a message and forget that everyone on the list will receive and potentially read their note. This can have embarrassing consequences.

MIME Type

This stands for Multipurpose Internet Mail Extension. It is a standard used to identify files by their extension or suffix. Applications, like your e-mail client, are said to be MIME conpliant when they can decode MIME suffixes. (See MOV, MPG, PDF.)

Mirror site

Some sites on the Internet are very popular and under heavy demand by the viewing public, and potentially overloaded with traffic. Mirror sites are exact copies of the original site that help to distribute the traffic load, increasing efficiencies in delivering information.

Modem

A modem is a device used to send and receive information across a phone line by your computer. Computers speak digital and telephones speak analog. Essentially, a modem is a translator. Modems are only one kind of device available for connecting your computer to the outside world. Two other methods becoming more common for home use are ISDN and cable.

MOV

This stands for movie. It is a file extension for animations and videos in the QuickTime file format.

MPG/MPEG

This stands for Motion Picture Experts Group. It is a format for both digital audio and digital video files.

Multimedia

As a general definition, multimedia is the presentation of information by multiple media formats, such as words, images, and sound. Today, it's more commonly used to refer to presentations employing heavy use of computer technology.

Nettiquette

This is a word created to mean *Network Etiquette*. It is a general list of practices and suggestions to help preserve the peace on the Internet. (See Flame.)

Newsgroup

This is the name for the discussion groups that can be on the *Usenet*. Not all news-groups are accessible through the Internet. Some are accessible only through a modem connection. (See *Usenet*.)

Pathname

This is a convention for describing or outlining the location of a file or directory on a host computer. A URL is typically composed of several elements in addition to the pathname. For example, in this URL: *http://www.prenhall.com/pubguide/index.html,* *http://* describes the protocol for a Web server, *www.prenhall.com* is the name of the host or server, */pubguide/* is the pathname, and *index.html* is the file name.

PDF

This stands for Portable Document Format. It is a file format that allows authors to distribute formatted, high-resolution documents across the Internet. A free viewer, Adobe Acrobat Reader, is required to view PDF documents.

Plug-in

This is a resource or program that can be added to a browser to extend it function and capabilities.

QuickTime (QT)

This is a file format developed by Apple Computer so that compters can play digital audio, animation, and video files. (See MOV, MPG.)

Robots

This is an automated program used to search and explore the Internet. Some popular search engines use these programs.

Search Engine

This is an online service or utility that enables users to query and search the Internet for user defined information. They are typically free services to the user. (See Robot.)

Search String

This is a logical collection of terms or phrases used to describe a search request. Some search engines enable the user to define strings with Boolean cues such as AND, NOT, or OR. (See Search Engine.)

Server

This is a software program used to provide, or serve, information to remote computers. Servers function in a Client-Server information exchange model. This term may also be loosely applied to the computer that is used to serve the information. (See Client.)

Shareware

This refers to software that is provided to the public on a try-before-you-buy basis. Shareware functions on the honor system. Once you've used it for a while, you are expected to pay a small fee. Two similar varieties of software are Freeware and Postcardware. Freeware is just that, free for your use and the owners of Postcardware simply ask you to send them a postcard to thank them for the product.

Shockwave

Shockwave is a plugin that allows Macromedia programs to be played on your Web browser. Many learning tools are beginning to be posted to the Web as Shockwave files. Visit Macromedia's Web site for more information and to download the plug-in (http://www.macromedia.com/). (See Plug-in.)

Signature

A signature is text that is automatically added to the bottom of electronic communications such as e-mail or newsgroup postings. A signature usually lists the name and general information about the person making the posting. Using a signature means that you don't have to repeatedly type your name and return information every time you send a note.

SPAM

This is the electronic version of junk mail. It also refers to the behavior of sending or posting a single note to numerous e-mail or newsgroup accounts. It is considered to be very bad nettiquette.

Stuff

This refers to the action of compressing a file with the Stuff-It program. This is a Macintosh format.

Table

This refers to a specific formatting element found in HTML pages. Tables are used on HTML documents to visually organize information.

Telnet

This is the process of remotely connecting and using a computer at a distant location.

Thread

This describes a linked series of newsgroup postings. It represents a conversation stream. Messages posted on active newsgroups are likely to spur numerous replies each of which can spin off into an independent conversation. The nature of newsgroups allows a reader to move forward or backward through a conversation as if moving along a string or thread.

Topic Drift

This describes the phenomena observed in many online conversations, typically chat or newsgroup, where the conversation will drift or change from the original posting.

Upload

This is the process of moving or transferring a document, file, or program from one computer to another computer.

URL

This is an abbreviation for Universal Resource Locator. In its basic sense, it is an address used by people on the Internet to locate documents. URLs have a common format that describes the protocol for information transfer, the host computer address, the path to the desired file, and the name of the file requested.

Usenet

This is a worldwide system of discussion groups, also called newsgroups. There are many thousands of newsgroups, but only some of these are accessible from the Internet.

User Name

This refers to an ID used as identification on a computer or network. It is a string of alphanumeric characters that may or may not have any resemblance to a user's real name.

Viewer

This is a program used to view data files within or outside a browser. (See External Viewer Application.)

Virtual Reality (VR)

This is a simulation of three-dimensional space on the computer. (See VRML.)

VRML

This stands for Virtual Reality Markup Language. It was developed to allow the creation of virtual reality worlds. Your browser may need a specific plug-in to view VRML pages.

WAV

This stands for Waveform sound format. It is a Microsoft Corporation format for encoding sound files.

Web (WWW)

This stands for the World Wide Web. When loosely applied, this term refers to the Internet and all of its associated incarnations, including Gopher, FTP, HTTP, and others. More specifically, this term refers to a subset of the servers on the Internet that use HTTP to transfer hyperlinked document in a page-like format.

Webmaster

This is the general title given to the administrator of a Web server.

Web Page

This is a single file as viewed within a Web browser. Several Web pages linked together represent a Web site.